Fourth Edition

Postharvest Technology of Horticultural Crops

Extension Methods and Capacity Building

Lisa Kitinoja

Founder
The Postharvest Education Foundation

University of California
Agriculture and Natural Resources

Davis, California
Publication 21664

UNIVERSITY OF CALIFORNIA
Agriculture and Natural Resources

To order or obtain ANR publications and other products, visit the ANR Communication Services online catalog at http://anrcatalog.ucanr.edu/ or phone 1-800-994-8849. Direct inquiries to

University of California
Agriculture and Natural Resources
Communication Services
2801 Second Street
Davis, CA 95618
Telephone 1-800-994-8849
E-mail: anrcatalog@ucanr.edu

Fourth edition, 2022

Publication 21664

ISBN-13: 978-1-62711-067-9

Library of Congress Cataloging-in-Publication Data

Names: Kitinoja, Lisa, author. | Kader, Adel A., editor. | Thompson, James F., editor. | Saltveit, Mikal E. (Mikal Endre), 1944- editor.
Title: Postharvest technology of horticultural crops. Extension methods and capacity building / Lisa Kitinoja.
Other titles: Extension methods and capacity building | Publication (University of California (System). Division of Agriculture and Natural Resources) ; 21664.
Description: Fourth edition. | Davis, CA : University of California Agriculture and Natural Resources, 2022. | Series: Publication / University of California, Agriculture and Natural Resources ; 21664 | Technical editors: Adel Kader, James F. Thompson, and Mikal Saltveit. | Includes bibliographical references and index.
Identifiers: LCCN 2020010583 | ISBN 9781627110679 (paperback)
Subjects: LCSH: Horticultural crops--Postharvest technology. | Food crops--Postharvest technology. | Agricultural extension work. | Agricultural education.
Classification: LCC SB319.7 .K58 2022 | DDC 635.028/6--dc23
LC record available at https://lccn.loc.gov/2020010583

Design by Sandra Osterman.

This publication has been anonymously peer reviewed for technical accuracy by University of California scientists and other qualified professionals. This review process was managed by UC ANR Associate Editor for Postharvest Technology and Biology Mikal Saltveit.

11/22-SB/LC/SO

Contents

Series preface

About one-third of fresh produce harvested worldwide is lost at various points in the distribution system between production and consumption. While it is impossible and uneconomical to eliminate these losses completely, it is possible to reduce them by at least half and increase food availability for the rising world population. Minimizing postharvest losses is more sustainable and environmentally sound than compensating for the losses by increasing the land and resources devoted to food production. The strategies for reducing postharvest losses include selecting genotypes with improved postharvest quality and longer postharvest life, using production methods that optimize postharvest quality, and using postharvest handling procedures to maintain produce quality.

Great advances have been made in postharvest technology through cooperative research by plant scientists, engineers, and food scientists. This interdisciplinary research, coupled with effective extension programs, needs to continue—to further improve the quality, flavor, nutritional value, and safety of perishable produce.

The goal of the technical editors and the authors of this *Postharvest Technology of Horticultural Crops* series is to describe the current scientific understanding of quantitative and qualitative losses in fresh-market vegetables, fruit, tree nuts, and flowers. The emphasis is on commercially available technologies that have been placed into practice to minimize losses and maintain food safety. Our perspective is based on the postharvest technology used in California and the United States. However, the principles discussed are applicable to fresh horticultural crops produced anywhere in the world. The information will be useful to people who want to learn about or are involved in growing, harvesting, packaging, transporting, or marketing perishable produce.

The *Postharvest Technology of Horticultural Crops* series is the outcome of a syllabus that was developed for a short course initiated in 1979 and offered annually since then in two modes: as a regular UC Davis course (Plant Science 196) for advanced undergraduate and graduate students and as a short course organized by the UC Davis Postharvest Center for participants who are not current UC Davis students. The latter group includes research and extension workers, consultants, quality and safety inspectors, packers, storage operators, transportation company representatives, and other professionals involved in postharvest handling of fresh horticultural perishables.

The first (1985), second (1992), and third (2002) editions of this series were published as single books and were well received, widely distributed, and used throughout the world. The amount of information on postharvest handling has continued to expand, and so this fourth edition has been converted into a series of ten books, each covering a separate area of postharvest technology.

We are indebted to our former colleagues Robert F. Kasmire, F. Gordon Mitchell, and Noel F. Sommer, who contributed greatly to the previous editions. Dr. Adel A. Kader initiated this fourth edition but was unable to see the conclusion of the publication because of his untimely death. We dedicate this series to him and have kept him as the lead technical editor on the series. He devoted his professional career to the cause of improving the postharvest handling of perishables. His enthusiasm, creativity, and unflagging commitment to his profession were an inspiration to us and to postharvest professionals around the world. We are indebted to Dr. Kader for his passion for bringing more and better-quality fruit and vegetables to the world's consumers.

On behalf of the authors, we also wish to thank all those who assisted in developing this series, especially Mary Reed, office manager of the Postharvest Center. We also thank the staff of UC Agriculture and Natural Resources Communication Services who participated in the production of this series.

James F. Thompson, Mikal Saltveit
Technical Editors

Postharvest Technology of Horticultural Crops
Extension Methods and Capacity Building

Introduction

An enormous amount of technically useful postharvest technology is described in the *Postharvest Technology of Horticultural Crops* series and other publications, and new or improved technologies for handling and marketing horticultural crops are regularly reported in technical and trade journals. Throughout the United States and abroad, postharvest extension specialists at universities and experiment stations recommend potentially useful postharvest handling practices in newsletters, extension publications, videos, and via the internet. Extension agents, farm advisors, postharvest consultants, and agricultural development personnel must be able to identify technologies that are cost-effective, feasible, and appropriate for their clients, as well as acceptable to consumers. With all the postharvest technologies from which to choose, how can extension professionals help their clientele determine whether any given technology will solve the problem at hand?

Identifying appropriate-scale postharvest technologies

Unfortunately, it is all too common to hear of technical fixes that caused more problems than they solved. During the 1970s and 1980s, postharvest losses were often attributed to a lack of storage facilities on-farm and in wholesale markets. Yet when large-scale commercial storage and marketing facilities were constructed they sat unused in many parts of the world. Recent examples of this problem come from Indonesia and Egypt, where large, government-funded packinghouses were never connected to power sources, and in Senegal, where a new wholesale market built opposite the existing market still sits empty, due mostly to vendors' unwillingness to pay higher fees for their booth spaces (fig. 1). The increasing use of participatory extension practices has helped to reduce such problems (Kitinoja 2010; Adhikarya 1994; Chambers et al. 1989; Chambers 1991; Cernea 1991a, 1991b), yet still during the late 1990s and 2000s various project evaluations and studies determined that the lack of use of new postharvest practices and structures in Africa,

Figure 1. Large, modern wholesale market space in a traditional Senegal wet market remains empty due to the high cost of booth rentals. *Photo:* Lisa Kitinoja.

Figure 2. Banana transport vehicle loading in India, where the fruit undergoes very rough handling. *Photo:* iStock.com/Dymov.

Figure 3. Tomatoes handled and packed in traditional baskets (known as tengas) in Tanzania. *Photo:* Janne Remmy.

India, and Latin America was due to their relatively high costs, inconvenient locations, management problems, or perceived security issues.

In the United States, consumers are often willing to pay more for produce that has been handled in what is considered a safe manner (through avoidance of pesticides, for example, or participation by farms and packinghouses in food safety programs). Extension professionals, by becoming aware of a host of possible factors that can affect whether a postharvest technology is cost-effective, feasible, and appropriate for the client, as well as culturally acceptable to consumers, can assist in the transfer of scale-appropriate information that is truly useful to their clientele.

Factors affecting adoption of postharvest technologies

Casual observation of or conversation with postharvest handlers and marketers of horticultural crops provides a wealth of examples of the typical practices and conditions that contribute to postharvest losses and quality problems in the United States and in other countries. Detailed studies of more than thirty commodity systems undertaken in sub-Saharan Africa and India during 2009 and 2010 for the Bill & Melinda Gates Foundation identified an enormous array of issues and poor practices (Kitinoja 2010; Saran et al. 2012). Use of poor-quality packages, rough handling, and lack of temperature management were found to result in very high levels of postharvest losses (figs. 2 and 3). The technical literature on postharvest handling of horticultural commodities documents the improper use of various practices as well as the use of traditional practices that appear to be counterproductive. Recent studies have documented that physical postharvest losses of fruit, vegetables, and cut flowers can reach 20 to 30 percent in the United States and often are as high as 50 percent in developing countries (Gustavsson et al. 2011; Lipinski et al. 2013). In addition, the loss of product value as quality declines during postharvest handling, storage, and distribution is an important source of economic losses. The following strategies have been recommended for reducing postharvest losses in developing countries: applying current knowledge to improve handling systems for horticultural perishables (especially packaging and cold-chain maintenance) and to ensure perishables' quality and safety; overcoming socioeconomic constraints, such as inadequacies of infrastructure, poor marketing systems, and weak R&D capacity; and encouraging consolidation and vertical integration among producers and marketers of horticultural crops (Kader 2005, 2010).

Common practices affecting postharvest losses, quality, and food safety

Many handlers unknowingly contribute to postharvest losses by using ineffective traditional practices or by not using certain practices known to reduce losses and help maintain produce quality and safety (table 1). The practices and conditions listed in table 1 have been observed within many horticultural production and fresh-marketing systems in both the United States and abroad. Each example is considered an improper practice since it has definite negative effects on fresh produce, leading either to increased waste and losses, quicker quality deterioration, or food safety problems.

Table 1. Common practices and conditions causing postharvest losses, reduced quality, and food safety risks

preharvest	• inadequate planning regarding planting and harvesting dates, or growing cultivars that mature when market prices are lowest • production of cultivars with high yields but short postharvest life or susceptibility to postharvest pests and diseases • use of poor-quality planting materials • overfertilization of vegetables with nitrogen • lack of use of calcium sprays for pome fruit • inappropriate irrigation practices • poor orchard and field sanitation, leading to latent infections and insect damage • lack of pruning, propping limbs, and/or thinning fruit, leading to small fruit with nonuniform maturation • lack of pest management (spraying for insect or fungal control, bagging produce susceptible to insect or bird damage)
harvest	• harvesting at improper maturity, leading to increased severity of storage disorders, lower eating quality (poor flavor and/or texture), failure to ripen, or excessive softening • use of rough and/or unsanitary field containers • harvesting during the hot hours of the day • rough handling, dropping or throwing produce, fingernail punctures • leaving long or sharp stems on harvested produce • long exposure to direct sun after harvesting • overpacking of field containers
curing	• lack of curing or improper curing of root and tuber crops before postharvest handling, packing, and storage • improper drying of bulb crops

Table 1. Common practices and conditions causing postharvest losses, reduced quality, and food safety risks, continued

packinghouse operations	• lack of proper sorting • lack of cleaning, washing, or sanitation • rough handling • improper trimming • misuse of postharvest treatments (over-waxing, inadequate chlorine in wash water, misuse of hot water dips for pest management) • use of inappropriate chemicals or misuse of registered compounds • long delays without cooling • lack of accepted and/or implemented quality grades or standards for commodities • lack of quality inspection
packing and packaging materials	• use of flimsy or rough packing containers • lack of liners in rough baskets or wooden crates • overuse of packing materials intended to cushion produce, causing interference with ventilation • containers designed without adequate ventilation • overloading containers • use of containers that are too large to provide adequate product protection • misuse of films for modified atmosphere packaging (MAP); overreliance on MAP versus appropriate temperature management
cooling	• in developing countries, general lack of the use of any methods of cooling during packing, transport, storage, or marketing of fruit or vegetables • in developed countries, use of inappropriate cooling methods, misuse of cooling methods, overcooling (chilling injury, freezing) • inadequate monitoring of temperature and chlorine levels in hydrocooler water
storage	• in developing countries, general lack of storage facilities on the farm or at wholesale or retail markets, lack of ventilation and cooling in existing on-farm facilities • poor sanitation and inadequate management of temperature and relative humidity (RH) in larger-scale storages • overloading of cold stores • stacking produce too high for container strength • mixing lots of produce with different temperature or RH requirements • lack of regular inspections for pest problems and temperature or RH management
transportation	• overloading vehicles • use of bulk transport or poor-quality packages, leading to compression damage • lack of adequate ventilation during transport • lack of air suspension on transport vehicles • rough handling during loading • lack of cooling during delays • ethylene damage or chilling injury resulting from transporting mixed loads

Table 1. Common practices and conditions causing postharvest losses, reduced quality, and food safety risks, continued

destination handling	• rough handling during unloading
	• lack of sorting, poor sanitation, improper disposal of culls
	• improper degreening of citrus crops and misuse of ripening practices
	• lack of protection from direct sun during direct marketing
	• in developing countries and farmers' markets, open horticultural markets exposed to sun, wind, dust, and rain
	• in developed countries, overcooling in supermarket displays of produce susceptible to chilling injury

Most of these improper practices and conditions cannot be labeled "technical problems," and they cannot be solved by initiating new research projects or simply by extending existing well-proven technical information on practices such as shade provision, improved packages, or low-cost cool-storage methods. Plenty of practical information on postharvest technologies suitable for smaller-scale producers, farmers' marketers in the United States, and smallholders in developing countries is already available at little or no cost (Cantwell 2007; Kader and Rolle 2004; Kitinoja and Kader 2003; Kitinoja and Thompson 2010; Lipinski et al. 2013; Patil 2010; FAO 1985, 1989a; Winrock International 2009). Yet often, postharvest losses take time to develop, so the specific cause of quality problems may not be fully understood by produce handlers along the value chain, and responsibility for the problem is easily passed along with the ownership of the product as it moves from one person to another. Other times, the handler may deliberately choose not to use a practice known to protect produce from losses because of its perceived cost or because consumers regard the practice as undesirable. On occasion, a lack of access to needed supplies, packages, reliable market information, or other infrastructural problems may make changes in handling impractical (Manalili et al. 2011).

Postharvest losses and changes in quality affect both the volume and perceived value of produce as it moves from the field to its final destination market, and any changes in practices also have an effect on final quality and related losses. Part of any potential technical solution, therefore, is a consideration of the socioeconomic, cultural, and institutional constraints facing growers, handlers, and marketers when they attempt to make changes in the way they handle and market horticultural crops.

Information gaps

Continuing use of some of these common improper practices can be traced to information gaps. Unless careful records have been kept of the handling practices undergone by a commodity between harvest and the final destination market, it can be difficult to determine the source of losses and quality problems. Even when fresh produce is moving through a simple direct-marketing channel from grower or shipper to retailer, many people are involved in the postharvest handling, transport, and marketing of it. The effects of poor postharvest handling practices are cumulative, and losses can be caused by one or more events. Perceptions of low levels of losses at any one point in the chain later may prove to be incorrect when the entire chain is considered and all the losses are added together. Identifying appropriate technical solutions can be impossible if the final handler is just guessing about when and where the cause of losses occurred and who may be responsible. This task becomes even more difficult if the extension agent is unfamiliar with the location, crop, language, or cultural habits of the community.

Once the extension professional helps clients close these information gaps, it is often found that a simple change can solve a major problem. For example, quality problems at final destination could be traced to a poorly designed package or a packinghouse cooling system that is being overloaded, causing internal temperatures of produce to be higher than desired when produce is being transported.

A variety of methods of postharvest loss assessment can be used to pinpoint a problem's source and to identify potential constraints to changing handling practices. Most involve direct observation of handling practices and interviewing key individuals regarding their standard postharvest practices. The United Nations Food and Agriculture Organization (FAO) has published loss assessment manuals (FAO 1985, 1989b) for various commodities that focus on measuring physical losses (changes in weight or quantity of produce) and losses in value (quality changes or decrease in market price per unit). Any method used for loss assessment must attempt to understand postharvest losses within the context of the whole system of production, handling, and marketing of the commodity in question, since what is considered a loss varies by culture and economic situation.

Postharvest systems research (PSR) is a loss assessment method that follows a given lot of produce from harvest through packing, storage, and transport to the processing facility or distribution center, all the while measuring changes in quantity and quality attributes (Shewfelt and Prussia 1993). Implementation of PSR requires the long-term use of a vehicle equipped with scientific instruments for quality assessment as well as sleeping accommodations for the researcher(s), since the many steps of postharvest handling may take days or weeks. To date, this method has been used primarily for applied research studies in horticulture at the University of Georgia. Another systems approach, originally developed by Jerry LaGra (LaGra 1990), is a practical team-based method. Used in fieldwork worldwide, it is known as Commodity Systems Assessment Methodology (CSAM). This manual was updated (LaGra et al. 2016) based on studies undertaken by the World Food Logistics Organization that used a modified CSAM to determine the causes and sources of postharvest losses for many different fruit and vegetable crops in Africa and South Asia (Kitinoja 2010). Loss assessment using CSAM is within the capabilities of most extension services, with its relatively low cost and its focus on face-to-face information gathering. Value chain analysis is the latest assessment method to gain a following in the European Union for agrifood market development (Vermeulen et al. 2008) and linking farmers to markets in Africa (Kitinoja 2010), although it is better known for analyzing manufacturing systems where the products are not usually perishable in nature.

Regardless of the methods used, postharvest loss assessment studies have provided information that suggests there are a variety of commonly occurring incentives and barriers to the adoption of new postharvest practices. Table 2 lists many factors that may affect the implementation of postharvest technology, some of which are discussed in detail in the following sections of this volume.

Socioeconomic factors

The potential for changing most improper handling practices is affected by a wide range of socioeconomic factors. In general, socioeconomic factors have to do with costs and benefits, whether actual, perceived, short-term or long-term. Some of the many examples are listed in the first section of table 3.

Growers of horticultural crops in many countries are tempted to harvest earlier than their neighbors (before adequate maturity) to take advantage of the high market prices that can be obtained early in the harvest season. While profits may be higher for the grower, the postharvest life and quality of the produce is greatly reduced, leading to higher postharvest losses and making handling and marketing more difficult.

In many countries, although curing is known to extend the postharvest life of root and tuber crops, curing can be viewed as too costly and unnecessary by growers intending to market their produce immediately after harvest. Yet if marketing is delayed for any reason, or uncured sweet potatoes are placed into

Table 2. Incentives and barriers to implementation of changes in postharvest technology

identifying sources of losses	• difficulties in determining amount and causes of losses • lack of knowledge about where and when losses occur and who is responsible • availability of training in loss assessment methods • availability of resources (money, time, equipment) for loss assessment
identifying possible solutions (technology)	• availability of scale-appropriate applied research results • profitability of postharvest technology (costs versus benefits) • governmental regulations, price supports or controls, incentives or disincentives • existing fee structures for common packages, transport, or storage services • technology fit with existing commodity system (handling methods, distribution system) • technology fit with local skills of handlers
postharvest equipment, supplies, fuel	• local availability of supplies, spare parts, tools, and equipment • costs and availability of materials, fuel, maintenance • reliability of supply of raw materials (for packing or processing) • reliability of supply of electricity • environmental concerns (is the technology reusable/recyclable?)
social or cultural norms	• cultural barriers (gender, religion, local customs, traditions, beliefs) • competing responsibilities (i.e., political or community service, social obligations, water bearing, child care, home-based tasks) • consumer demands and preferences (definitions of quality vary by culture) • requirement of produce "losses" for other purposes (i.e., animal feed, poverty relief programs) • environmental concerns (a new technology may create pollution) • effects on jobs (may displace workers) • existence of grades and standards for produce
credit and loans	• availability of credit • collateral requirements for loans • prevailing market interest rates
postharvest labor/who will implement the change?	• seasonal changes in labor supply, labor shortages • availability of trained personnel for harvesting, packing, and skilled jobs • motivation and loyalty of the workforce
training, education, and instruction	• availability of resources for postharvest educational and training programs • availability of effective extension services (public or private) • differences in access to existing services (women versus men, small- versus large-scale enterprises, farmers' marketing cash crops versus food crops) • extension workers' level of knowledge and skills in postharvest handling and/or food processing

Table 2. Incentives and barriers to implementation of changes in postharvest technology, continued

supporting infrastructure	• existence and condition of roads • availability of cooling, storage, transportation, and marketing facilities • location and size of existing facilities • access to existing facilities • institutional linkages (i.e., communication between research and extension) • government investment priorities in facilitating services • budget for maintenance of existing infrastructure • availability of staff with management skills for running facilities • condition of communications systems (telephone, mail, fax, e-mail services)
marketing system	• availability of marketing options and alternative outlets • direct sales versus consignment (who is responsible for losses? who suffers financially?) • degree of cooperation and coordination among buyers and/or sellers • establishment and viability of marketing cooperatives • reliability of record keeping to help trace problems back to sources
market information	• availability of resources for collecting and analyzing data • reliability of information • timeliness of information

storage by the buyer, postharvest losses due to water loss and decay can be severe.

Sometimes existing circumstances contribute to the continuing use of a detrimental postharvest practice. In the Caribbean and much of Africa, freight charges for produce are set by the number of containers rather than by the containers' weight or size. This policy provides an incentive for handlers to use the largest possible container in order to reduce their shipping costs, regardless of the lack of protection offered by such packages to the produce within. Recent examples come from Ghana and Tanzania, where cabbage is loaded into jute sacks that have been expanded beyond their intended size so traders can avoid paying freight for additional packages (fig. 4), and Ghana, where tomatoes are packed in enormous 80-kilogram wooden crates (fig. 5).

Even when the overloading of containers and transport vehicles is a well-known cause of losses due to compression damage and reduced ventilation, overloading is common because it reduces immediate handling and transportation costs. While handlers may explain that they use these practices because the number of vehicles is limited or because handling a few very large packages is quicker than hand-stacking many small packages, these practices always compromise quality later in the postharvest chain.

Large-scale growers or shippers may use a particular practice because it is cost-effective, and losses of a certain magnitude may be deemed acceptable (Stuart 2009). Redesigning and producing a new package for specialty fruit, for example, might cost more than the new package saves in reduced losses. Profit-motivated decisions, such as choosing to use less packaging to maximize individual return on investment, can increase postharvest losses later in the commodity system. Yet any suggestion that packages be modified to better suit a commodity's needs runs into serious opposition, since the added cost is borne by the packer but the benefits of reduced losses are gained by others who own the product later in the handling chain.

Sometimes the reasons for the choice of a given postharvest practice are complex. While

Figure 4. Packing cabbage into expanded sacks in Tanzania. *Photo:* Bertha Mjawa.

Figure 5. Tomatoes arrive in Kumasi, Ghana, from Burkina Faso, shipped in enormous wooden crates. *Photo:* Adel A. Kader.

field packing is known to reduce losses by decreasing the number of times produce is handled, choosing to field-pack vegetables or to use a packinghouse depends on many factors. The practical decision may be based on the amount of new investment required, the land, equipment, and facilities already owned or available, and the expertise of managers and packers. In the United States, operations often face a dilemma when, as they prepare for a postharvest handling practice such as grading, they must choose whether to rely on manual labor, which can provide more jobs, or on machinery designed to enhance efficiency. Which choice is better for a given operation and its community depends on the philosophy of the owners, the cost of equipment, the expected returns, the availability of trained labor, and the level of local unemployment.

In general, investing in postharvest technology has been demonstrated to be highly cost-effective (Kader and Rolle 2004; Kader 2006; Kitinoja 2010), but extension educators must always consider socioeconomic factors whenever assessing the suitability of postharvest technology for their clientele. Small-scale handlers may not be able to afford to implement well-known technical practices such as sorting, cooling, or improved packaging on their own without reducing profits. For small farmers, whether to use precooling and cool-storage methods depends on the type of crop being handled and on degree of access to the power or water needed for the cooling technology all along the value chain (Kitinoja and Thompson 2010). If investing in the technology requires the use of credit, prevailing interest rates or access to collateral can affect decisions about whether to change handling practices. The cost-effectiveness of investing in a postharvest technology will always be a primary factor in whether it will be considered appropriate by an individual, a group of farmers, a company, or cooperating marketers.

Cultural factors

In general, cultural factors affecting the implementation of changes in postharvest technology have to do with beliefs, preferences, traditions, or cultural norms. The middle section of table 3 lists cultural factors that may interfere with making changes in postharvest technology.

Recommended postharvest technologies must fit into the existing cultural environment. For example, technical factors may suggest packaging a product in a certain way to reduce losses and protect quality, while consumers prefer to make their own selection from a bulk display. Some consumer groups may prefer processed or fresh-cut produce and be willing to pay more for the convenience offered by these products, while other consumers will avoid them due to perceived nutritional or health issues.

Beliefs regarding the environment may influence consumers' food choices and can affect how they perceive produce wrapped in plastic or packaged in foam containers. Handlers may find that using recycled materials in packaging may be profitable from a monetary as well as a goodwill point of view as consumers look for "green" products. Even though concerns over the use of pesticides, food additives such as waxes or colorings, or postharvest treatments such as gamma irradiation are considered by scientists and the health care profession to be largely irrational, their use may make produce less valuable to specific groups of consumers. Market research studies undertaken by extension professionals can help identify key consumer preferences and the level of demand for fresh produce in various sorts of packages.

Sometimes local customs or consumer preferences lead directly to quality problems. For example, it is often recommended that harvests be performed early in the morning in order to reduce the heat load on produce and make initial cooling faster and less expensive. Vegetables in West Africa, however, are often harvested at midmorning and endure the heat of the day while awaiting transport from the field. In this case, the women who harvest the vegetables cannot come to the fields earlier, since family responsibilities such as water bearing and cooking must be given first priority. Another example of the effect of a cultural factor on postharvest handling occurs in New Delhi, India, where fresh produce is sold by weight. Leaving a few protective outer leaves

on cole crops or sprinkling leafy greens to prevent wilting is viewed as "cheating" since these practices add weight to the produce, so vendors tend to avoid these practices and produce is wasted.

Other cultural factors that can affect whether handlers adopt changes in postharvest technology include religious traditions, gender barriers, the local definition of losses, and traditional secondary uses for low-quality produce (e.g., animal feed, food banks). If agroprocessing is to play a role in reducing food losses, recommended technologies must result in high-quality, healthful foods that consumers find good-tasting and easy to prepare. Extension professionals must be aware of these cultural factors in order to best identify scale-appropriate postharvest technology and before attempting to develop educational programs targeting specific clientele.

Institutional factors

The adoption of postharvest technology can be affected by existing laws or regulations providing incentives or disincentives, as well as by facilitating services provided or not provided by governments, universities, and the private sector. The third column of table 3 provides examples of various institutional factors that may affect whether people make changes in postharvest technology; additional challenges and opportunities are presented and discussed in Kitinoja et al. (2011) and Kader et al. (2012).

Institutional factors include public-sector policies such as commodity price supports or controls, land tenure and property rights, and regulations regarding the use of pesticides and acceptable levels of residues. Secondary institutions can become involved when, for example, fees are charged as part of mandatory marketing orders. The status of these policies and regulations within a state or country may affect whether someone is willing or able to make individual changes in their postharvest practices. For example, the decision to provide price supports may increase production, which may lead to the requirement for more supplies for postharvest treatments, packaging, and storage facilities. If any of these are inadequate,

Table 3. Factors affecting implementation of changes in postharvest technology

Socioeconomic factors	Cultural factors	Institutional factors
sun-drying vegetables directly on the soil (dust, insects, molds) because of high cost of plastic sheeting (Morocco; Kitinoja 1996)	harvesting onions very early, leading to high losses—but responding to consumer demand for the flavor of early onions (Chad; Kitinoja 1992)	government fees for using potato storage facilities so low that the facilities sat empty (operators would have lost money by running them) (UP, India; Kitinoja 1995a)
very early harvesting (before maturity) to take advantage of higher market prices at beginning of season (India; Reid et al. 1997) (Morocco; Kitinoja 1996)	health and environmental concerns (irradiation, waxes, pesticides, nonrecyclable packaging materials) (USA; Krimsky and Plough 1988)	women being unable to own property and thus often having no collateral to offer in return for loans (Africa; Madeley 1987)
inter-island freight rates charged by unit rather than by weight or volume of produce (Caribbean; Schurr 1988)	washing produce associated with the use of toxic chemicals, specifically pesticides (lowers perceived value) (Morocco; Kitinoja 1996)	poor roads; lack of loading docks, cooling and storage facilities, telephone services; inadequate power supply (Costa Rica; Breslin 1996)
overloading transport vehicles and packing containers because of limited availability (LDCs; FAO 1989b)	canned goods actively disliked by consumers (fear of contamination, preference for fresh, in-season foods) (India; Reid et al. 1997)	bureaucratic delays resulting in government facilities for forced-air cooling and refrigerated transport remaining unused for years (Punjab, India; Reid et al. 1997)
using less, or lower-quality, packaging materials to cut costs (worldwide; LaGra 1990)	late-morning harvesting due to women's responsibilities in child care and water bearing (Senegal; Kitinoja 1995a)	market information collected but not disseminated to potential users in a timely manner (LDCs; FAO 1989b)
very high interest rates for loans (LDCs; FAO 1989b)	custom whereby only men operate machines, so any mechanization may take women out of the postharvest system (Africa; Chinsman and Fiagan 1987)	lack of standardized grades and inspection services (LDCs; LaGra 1990)
lack of financial opportunity to make investments in onion storage facilities (Chad; Yarnell 1990)	male extension agents' difficulty working with women farmers and marketers in Islamic communities (Senegal; Kitinoja 1995a) (Africa; IFPRI 1995)	government horticultural research and extension services focusing solely on production (India; Kitinoja 1995a)
sprinkling leafy greens to reduce wilting considered "cheating" since produce is sold by weight (India; Kitinoja 1995a)	farmers' preference for storing crops inside their homes due to security concerns (leaving large-scale storages unused) (Peru; Rhoades 1984)	offering new technologies with little or no consultation with locals (worldwide; Wiggins 1994) (Ghana; Compton 1997)

lacking, or too expensive, postharvest losses may increase.

Also of concern are the availability and condition of facilitating services such as the physical infrastructure (e.g., roads and storage facilities), electricity supply, quality control and inspection services, availability of credit or loans, extension systems, and communications and market information systems. When services related to postharvest handling are nonexistent, skewed to serve certain groups over others, or poorly managed, adoption of improved postharvest technology can be negatively affected. For example, in many countries, market information is collected but not analyzed and hence goes unused. The FAO suggests limiting the scope of a market information system to a few commodities in major markets, then aiming to collect, analyze, and disseminate information (on prices, supply, and movements of produce) to users immediately (on the same day or at latest the next day). And while many of the people involved in horticultural production, handling, and marketing are women, most extension agents are men. An International Food Policy Research Institute report noted that "if more extension agents and agricultural research scientists were women, extension services and agricultural technologies could be made more appropriate to female farmers. The representation of women in extension services is 'miniscule'" (IFPRI 1995), and this is still true today. Training specialists have suggested that home economics agents (also known as family resource or home advisors) may be better suited to working directly with women than are most agricultural agents, since contacts between home economics agents and rural women have already been established.

Extension of postharvest information

Whenever extension efforts are undertaken in postharvest technology, it is important that the appropriate audience be targeted. For example, harvesters need to learn about maturity indices, while transport operators need information on temperature management during loading and shipping. California, with its excellent array of facilitating services, including the University of California (UC) Cooperative Extension, serves as a model for successful cooperation between the government, university research, extension services, and the horticultural industry. In many regions of the world, however, poorly developed infrastructure hinders the movement of produce and market information (Kitinoja et al. 2011). The lack of linkages between agencies or a lack of resources for extension work and agricultural education leaves handlers without basic knowledge about causes of losses and with a shortage of skills related to improved postharvest handling practices. Postharvest educational needs assessments undertaken in many countries point to a wide range of training requirements before local handlers are ready to make improvements in postharvest technology within their operations.

Extension workers worldwide have an opportunity to extend appropriate postharvest technology to a variety of clientele who can then use recommended practices to reduce losses, maintain produce quality, and increase profits. The final decision about whether to adopt any given improved postharvest technology will be made on an individual basis, and will depend on many factors that affect the technology's ultimate usefulness. The more information extension workers have about the postharvest problem, the client's operation, the local situation, and the local costs and expected benefits, the better they will be able to identify effective solutions. Involvement of clientele in postharvest loss assessment can assist extension workers in pinpointing the sources and causes of losses and in identifying potential constraints to making the changes required to solve the problem. Commodity systems assessments can help to identify who (i.e., men, women, growers, traders, retailers) needs what kind of postharvest information to solve the problem. Market research can help identify consumer preferences and demand characteristics that affect the feasibility of certain solutions. Only then can effective postharvest extension programs be developed to meet the needs of local clientele. Specific postharvest handling recommendations, therefore, are best made by local development agents or extension workers who have had a chance to review the entire commodity system in question, gather information from those involved with postharvest handling, and become familiar with the local cultural, institutional, and socioeconomic environment.

There is a great need for effective local extension services in postharvest technology to help solve the industry's problems and to disseminate research results and other information useful to growers of fresh-market horticultural crops and those involved in transportation and marketing. The objective of extension is to create a two-way link between information providers and information seekers. Extension professionals must ensure that developers of postharvest technology are made aware of the priority problems of large-, medium-, and small-scale clientele, and that potential users of postharvest information find the technology accessible, easy to understand, and suited to their needs and constraints. To fulfill this objective, extension personnel must be highly motivated, well trained, and equipped with the latest resource materials, reliable transportation, and communications technology. They must be located where they can readily and effectively interact with researchers and those involved in the preparation, shipping, distribution, and marketing of horticultural crops.

The objectives of an extension postharvest program are to improve the quality and value of horticultural crops available to consumers, reduce marketing losses of horticultural crops, and improve marketing efficiency. All of these relate to the profit motive. Most growers and handlers do not make changes in their postharvest operations just because they want to reduce losses—the changes they make must also lead to improved profits. Specific extension objectives will focus on solving a particular problem affecting one commodity or a group of related commodities in a specific location or commodity system. These objectives could be focused on any aspect of postharvest extension, from ensuring quality and food safety at harvest to helping clients meet

consumer preferences and the level of demand in new markets (table 4).

The clientele of the typical regional or national postharvest extension specialist consists of local extension workers (farm advisors, county agents, agricultural technicians, or village-level workers) and industry personnel involved in preparation, shipping, and distribution of fresh-market horticultural products. The clientele may also include students from universities or agricultural colleges with whom the extension specialist comes into contact through teaching, guidance, and counseling responsibilities. The local extension worker's clientele consists primarily of farmers and horticultural industry personnel within a specific county, district, or region.

To plan any postharvest extension program, one must first identify a specific problem and then identify and understand the needs of the clientele being addressed. Sometimes the audience's composition, background, and distribution are major constraints. The fresh-produce shipping, marketing, and distribution industry is very heterogeneous and widespread, often including handlers at distant points in the same country or in more than one country. Farmworkers or market intermediaries may be less educated than average, or extension programs might target clientele who speak a language different from the mainstream. Initially, clientele may know little, if anything, about extension's role and objectives and may be suspicious of any attempts to convince them to change their handling practices, techniques, or facilities. Outside the United States, the reputation of extension has been tarnished by poorly designed extension systems that failed to serve their clientele. Indeed, extension work is commonly termed "outreach" to avoid the negative connotations associated with the word "extension." To gain the attention and confidence of a reluctant clientele, a program must first demonstrate that it can benefit them, most often economically.

Table 4. Examples of extension postharvest program objectives, extension messages, and related research topics for postharvest extension programs

Postharvest activity	Extension objective	Extension message	Related applied research topics
crop selection	ensure that farmers grow crops appropriate to market conditions and demand	introduction of "new" crops with market potential	prices, quality, yields, returns, and marketing strategies
harvesting	ensure that produce is harvested at the time or quality required by the market	correct time to harvest crops for marketing, and correct quality level	adaptation of technology to meet time or market requirements
preparation for market	maximize value to growers	on-farm cleaning, trimming, and selection of produce for market	adaptation of technology to meet time or market requirements
grading	allow pooling of output and collective marketing; improve output quality	grading methods for horticultural products, advantages of selling graded produce	formulate and monitor standards for marketable produce
packaging	maximize returns to producers and handlers by minimizing damage	demonstrate packaging products for transport, storage, and sales	adapt traditional packaging materials to crop and market requirements
transport	ensure that produce reaches buyers without delay or loss	encourage improved handling and packaging procedures to reduce losses during transport	cost-effective means of transport
storage	reduce growers' dependence on intermediaries	how to determine when, where, what, and how to store	adaptation of storage procedures to local conditions and crops
sales/marketing	improve growers' bargaining position, reduce dependence	encourage collective marketing	market surveys, analyses of prices, and demand patterns and characteristics

Six steps for developing high-quality postharvest extension programs

After identifying and learning about the intended audience for a postharvest extension program, take the following helpful steps.

Step 1: Identify the postharvest problems to be targeted and work with stakeholders (planners, funding agencies, cooperators, clientele) to determine their priorities. A traditional educational needs assessment or commodity systems assessment, undertaken with the participation of representatives of the intended audience, can help identify key postharvest and marketing problems. It is not possible to develop extension educational programs for all problems at one time. Determine which are the most important for a given audience and which can be realistically resolved by providing educational information on postharvest principles and/or practices.

Step 2: Develop the long- and short-range (1-year) objectives of the extension program and the program's theory of action. The typical chain of program events is described by a model developed by Bennett (1979):

- *Inputs* and resources must be used to get the program started;

- *Activities* are then implemented to *involve* people in programs;

- Participants *react* to what they experience;

- Which leads to *changes* in *knowledge, skills, attitudes,* or *aspirations*;

- *Changes* in *practice*; and finally in

- *End results*.

For example, a long-range objective might be to improve the nutritional quality of fresh fruit and vegetables sold to consumers; a short-range objective might be to improve cooling practices and facilities used before transportation.

Step 3: Describe the specific postharvest principles, practices, and technologies to be offered during the extension program. Take into

consideration any known or suspected constraints that may influence the adoption of postharvest technology. Paying attention to the many factors discussed in the first part of this volume will help extension workers focus on those technologies that will be most appropriate for their clientele.

Step 4: Describe the extension methods that will be employed to reach the program's objectives. Indicate whether methods will be individual, group, mass media, or based on information technology. Identify information sources and available teaching materials. It is best to involve cooperators from the produce industry or related organizations in extension programs whenever possible, both to benefit from their expertise and to gain their support for program objectives.

Step 5: Determine the resources needed to conduct the program. Compare the workforce, equipment, and facilities that will be needed with those that are currently available and identify possible sources of funding, resource persons, tours, demonstration sites, and supplies.

Step 6: Develop a plan for evaluating the program during implementation in order to determine whether objectives are being met and how the program might be improved. Include the program's stakeholders in the evaluation to enhance the chances that the results of the evaluation will be used.

More information on the topics of educational needs assessment, commodity systems assessment, extension methods, and program evaluation can be found later in this volume. For more detailed information on the extension program planning process, see Blackburn (1994), Van der Ban and Hawkins (1996), and the website of the Postharvest Education Foundation, postharvest.org.

Extension methods

Many extension methods are available for conducting postharvest programs within any extension system. Successful extension work is an art as well as a science. The relative effectiveness of any method depends on the level of interest and voluntary involvement of postharvest researchers, public and private extension workers, and industry clients. Always, the work of extension is to build new links and strengthen existing links between information generators and information users. These links help make postharvest researchers aware of industry problems, perspectives, and constraints, and also help increase industry awareness of the problem-solving assistance available from scientists. The following is a discussion of some of the methods most commonly used in postharvest extension programs.

Applied or adaptive research

Applied research studies in postharvest technology seek to identify causes and magnitude of deterioration or losses and to develop and evaluate possible corrective measures. Adaptive research attempts to modify an existing technology to better fit the exact conditions in which it will be used in practice. While some research must be conducted in laboratories, and other research can be conducted in industry facilities, all research must use scientifically sound methods and procedures.

Collaborative and participatory research and development methods ensure that the postharvest information developed by extension specialists is useful and appropriate for their clientele. UC Davis postharvest faculty members are involved in many adaptive research projects, including the Horticultural Innovation Lab's projects in Rwanda, Tanzania, Uganda, and Zambia, where appropriate packaging materials, cool-storage structures made of indigenous materials, and small-scale food processing methods are being investigated (Kitinoja and Barrett 2015). The skills needed in applied research include patience and attention to detail, careful planning, and willingness to listen to cooperators and clientele. See Andrew and Hildebrand (1993) and Ewell (1990) for further information and case studies of successful applied research methods.

Public extension services face a unique challenge whenever they work with clientele associated with large, technically sophisticated companies involved in produce handling and marketing. Due to a competitive outlook that makes it undesirable to share with others the results of applied research studies, the private sector may be unwilling to cooperate fully with postharvest extension specialists.

Consultations

Requests for consultations most often originate with industry leaders or groups who want to improve their postharvest operations. Consultations are usually done in person, but they can also be handled via telephone, the mail, e-mail, or the internet and can involve individuals, companies, or other groups such as grower cooperatives. Consultations generally deal with specific subjects (e.g., decay problems or cooling methods) and individual problem diagnosis. Despite the high time requirement associated with individual consultations, they continue to be an important extension method. Extension workers can use the time they spend with individuals to obtain informal information on needs and feedback on prior or ongoing extension programs, and to build linkages with clientele.

Group meetings

Meetings with groups allow extension workers to present information to larger audiences than do consultations. Regularly scheduled group meetings give clientele the opportunity to feel connected and receive updates and progress reports on ongoing applied postharvest research and extension work. Group meetings also encourage direct audience participation and discussion. When postharvest researchers are invited to participate, clientele can offer direct feedback regarding the information presented, express their needs and concerns, and explain the constraints they face.

Reaching growers and large-scale shippers and retailers is relatively easy, but postharvest handlers and small-scale marketing intermediaries sometimes fall through the cracks. Intermediaries are mobile by nature and may work in various regions at different times of the year and sell in several markets. In many developing countries, women do the majority of food marketing, while men handle cash crops. Market intermediaries there are often illiterate and learn their jobs by apprenticeship with an elder. Typically, government agencies, universities, and extension services focus on growers (through ministries, departments, and colleges of agriculture) and wholesale or retail vendors (via ministries or departments of commerce or marketing), but no agency specifically targets market intermediaries. Successful extension of postharvest technology to traders will require methods that use a hands-on approach.

As is usually the case with extension activities, the language and socioeconomic status of the extension worker should match those of the clientele for best results. The use of informal and hands-on teaching methods whenever clients are illiterate helps increase participation and comprehension. When women are part of the target audience for postharvest extension programs, meeting times must be scheduled with the needs of women in mind. The locations selected for meetings must be accessible and provide some form of child care facilities. The marketplace can be an excellent location for many extension activities related to postharvest handling practices. As natural gathering places, wholesale and retail markets are excellent locations for information gathering, tours, and demonstrations of recommended postharvest technologies. Holding regular group meetings in local wholesale and retail marketplaces also helps reduce the transportation costs associated with extension work and with clientele participation.

Demonstrations

Extension workers can use demonstrations to show how to use a new practice, procedure, or facility or to illustrate the results of recommended postharvest technology (Barao 1992). Demonstrations are often used to extend the results of applied or adaptive research. A hands-on, or experiential, learning approach can enhance program results, since many people "learn by doing." Careful attention to the equipment, facilities, and visual aids used in demonstrations can increase their effectiveness.

The message of any given demonstration should be simple and clearly presented to participants. One example is the effect of various temperatures on the postharvest life of selected produce; another is the effect of chlorine on the vase life of roses. In a workshop conducted in Upper Egypt, two equal-weight sample bunches of unpackaged spinach were placed in an air-conditioned room and on a balcony outside in the heat, and weight loss was measured 1 hour later to demonstrate how much more water had been lost at the higher temperature. In a short course offered in India by UC postharvest specialists during the hot dry season, Punjabi flower producers were enthusiastic about their training when they witnessed a large vase of wilting roses seemingly brought back to life when a dash of chlorine bleach and lemon-lime soda was added to the vase solution.

An important component of any demonstration is a cost-benefit analysis, in which a comparison is made between the current handling practice and the new postharvest technology under consideration. A simple chart for comparing any two practices can be constructed and used to calculate the relative costs and the expected benefits (table 5). Costs include required equipment, labor, supplies, and power; benefits might include increased volume of produce due to lower losses, higher quality grade, or better market price. In the example shown in table 5, an investment in reusable plastic crates has a relatively high cost for smallholder farmers but an immediate positive return on investment, since losses are reduced from 30 to 5 percent and the market value of the produce is 25 percent higher than when it is transported in sacks. For more details on calculation methods, refer to Kitinoja (1999) and Kitinoja and Gorny (1999). For detailed examples of a variety of cost-benefit studies on the use of shade, improved packages, evaporative cool chambers, cold rooms with CoolBot technology, and small-scale food processing methods, refer to Kitinoja (2010, 2013).

There is a strong correlation between the characteristics of an innovation and its rate of adoption (Rogers 1995). Table 6 lists the characteristics that are known to be important. Postharvest technologies that have one or more of these five characteristics can

Table 5. Relative costs and expected benefits for the use of reusable plastic crates, assuming a harvest of 1,000 kg

Costs and benefits	Current practice	New practice
	Sacks for sweet peppers 25 kg	Reusable plastic crates plus fiberboard liners 12.5 kg
Costs		
40 sacks @ $0.50	$20.00	
80 crates @ $6.00		$480.00
Crate liners @ $0.10		$8.00
Relative cost	$20.00	$488.00
Recurring costs	$20.00	$8.00
Expected benefits		
% losses	30%	5%
Amount for sale	700 kg	950 kg
Value/kg	$1.00/kg	$1.25/kg
Total market value	$700.00	$1,187.50
Market value – costs =	$680.00	$699.50
Relative profit 1st load		+$19.50
Relative profit 2nd and subsequent loads		+$499.50

Note: Immediately profitable with first load. Once plastic crates have been paid for, the relative profit will rise to +$499.50 per load of 1,000 kg. Plastic crates can be used more than 100 times.

Table 6. Characteristics of an innovation that enhances adoption

Characteristic	Comments
cost	Does the innovation enable the client to achieve goals better or at lower cost than previously? Postharvest technology that is clearly cost-effective will be of most interest to potential users.
compatibility	Is the innovation compatible with sociocultural values and beliefs, with previously introduced ideas, and/or with clients' felt needs? Any new postharvest technology must not cause more problems than it solves.
complexity	Can the innovation be adopted without complex knowledge or skills? If the postharvest technology is difficult to understand or use, clients will be less likely to want to try it for themselves.
trialability	Can the client try the innovation on a small scale before deciding whether to make large-scale changes in practices? If a large investment is required before the user can see any results, the postharvest technology will remain a training exercise.
observability	Can the client see the effects of changes made by others who have adopted the innovation?

Source: Rogers 1995; Van der Ban and Hawkins 1996.

be demonstrated successfully in extension programs.

Short courses

Extension audiences can benefit from intensive, broad coverage of specific subjects in short courses, which can include classroom lectures, laboratory demonstrations, or tours. The Postharvest Technology Short Course offered annually in June by UC Davis is an example, with 5 days of classroom instruction followed by 5 days of tours. The University of the West Indies conducts short courses throughout the Caribbean on postharvest topics via their Continuing Education Program in Agricultural Technology. Other short courses are conducted by individual marketing firms for their own personnel or by trade association specialists. The UC Postharvest Center website, postharvest.ucdavis.edu, is an excellent source of information on current postharvest course offerings, and it also has links to other sites.

Short courses may run between 2 days and about 2 weeks. They may be used to refresh information that an audience has previously learned and to provide updated or new information on a subject. When the course meets for several days or more, the extension specialists can prepare demonstrations of the use of postharvest practices and clients can see for themselves the results of certain technologies.

Effective short courses require much planning, considerable professional involvement and input, proper facilities, and follow-up to evaluate their effectiveness. A good mix of extension methods should be used to maintain interest, and too much lecture-style presentation should be avoided. The use of visual media (Powerpoint presentation, videos, CD-ROMs or DVDs), group exercises, laboratory activities, and discussion helps keep participants involved. Printed materials, including a syllabus and a list of current references for further information, are generally provided to each participant in a short course.

Workshops

Workshops can improve the skills in postharvest technology of individuals or groups. For example, one might conduct a specific workshop on harvesting, cooling, or careful handling. Often, a workshop will focus on a single commodity and provide written materials as well as use visual aids and include discussions. Workshops can last for 1 or more hours and can meet only once or as a series over a period of time. Organized meetings and workshops in California have led to changes in practices in the apple industry, where growers have learned the importance of harvesting earlier and storage operators have learned to monitor and maintain low levels of carbon dioxide in storage. These simple changes in postharvest practices have resulted in a decreased incidence of internal browning in Fuji apples.

The scheduling of workshops can be a difficult issue, since clients may be too busy handling a product to attend during a time that is optimal for the extension worker to demonstrate pertinent postharvest handling practices for the product.

Study tours

Postharvest facilities and operations are often easier to understand after a well-run observational tour. Tours can be a part of a demonstration, short course, or workshop. The University of Florida's Horticultural Sciences department offers an annual 4-day Florida Postharvest Horticulture Institute and Industry Tour in the early spring, during which participants visit harvest, packing, shipping, and warehousing operations throughout south and central Florida. UC Davis offers an optional 5-day tour of California postharvest operations following its annual June Postharvest Technology Short Course.

A tour can be an effective way to introduce a new subject to an audience. Providing a brief outline of what participants might expect to see can help them focus their observations. Industry cooperation is essential for a successful tour, since participants will have many questions and will appreciate the opportunity for direct interaction with the owners or managers of the site. Tours require excellent planning and skills in logistics and communication to keep everyone on track and on time.

E-learning programs

A low-cost extension method used for training postharvest trainers and extension workers is an internet-based distance education program (also known as an e-learning program or MOOC). An example is the Global Postharvest E-learning Program, which is offered each year by the Postharvest Education Foundation, postharvest.org. Participants enroll in a year-long program that includes ten assignments (readings, fieldwork, and written reports) on commodity systems assessment, postharvest demonstration design, postharvest extension program planning, cost-benefit analyses, and design of a postharvest training and services center. Since 2012 more than 150 young horticultural professionals from twenty-eight countries have successfully completed this global e-learning program.

Publications

Publishing in professional journals, trade publications, newsletters, and other media outlets helps extension workers extend information to a greater, more distant audience than can be reached through personal contacts. Effective articles present information clearly, succinctly, and in an appealing manner; they are addressed to a specific audience for a specific purpose, and they do not impose the writer's personal bias on the information presented. A publication can be written in a variety of formats, depending on the intended audience. Results from specific studies or other relevant information are often published in two or more types of publications to reach a broader audience and achieve the maximum effect. For example, the scientific details of a study might be reported in a professional society journal and the semitechnical aspects published in a university or government report. A popular report of the study, showing its relevance to the postharvest industry, might be published in extension newsletters and trade publications. Following are examples of types of publications.

Technical and semitechnical articles. These include articles on applied research published in professional societies' publications and in university and government technical reports. They are written in a scientific style, primarily for the benefit of professional researchers and extension workers; few industry representatives read these publications.

Brief, single-subject guides. This type of publication addresses a single specific subject or development in a direct, simple style. The produce fact sheets published by the UC Postharvest Center, postharvest.ucdavis.edu/Commodity_Resources/Fact_Sheets/, are examples. These fact sheets, which provide recommendations for maintaining the postharvest quality of an individual commodity, are published in English, Spanish, French, and Arabic.

Progress reports. These publications extend current information on ongoing projects to cooperators, research sponsors, industry personnel, and others. An example might be results of a just-completed preliminary study on the effects of a questionable, currently used industry practice. They are usually brief reports, no more than a few pages long. Their main advantages are timeliness, brevity, and directness, and they keep interested persons informed about ongoing results.

Newsletters and quarterlies. These periodical publications extend information to broad audiences on a regular basis, typically four to six times a year. A good newsletter is an effective route for extension of brief, pertinent reports and articles to the postharvest industry and to fellow extension and research workers. The *UC Perishables Handling Quarterly* was an effective informational tool from its founding as a newsletter in 1962 and is now available online as a regular e-newsletter at postharvest.ucdavis.edu/E-News/. Issues usually contain a review article on a specific subject related to postharvest handling, brief articles on recent research results, book reviews, and a list of recent postharvest publications and reports. The University of Florida IFAS postharvest newsletter for industry, *Citrus Packinghouse Newsletter*, irrec.ifas.ufl.edu/postharvest/newsletter/index.shtml, is an effective source of information, as is a website maintained by the SAVE FOOD Initiative, fao.org/food-loss-reduction/news/en/. The horticultural industry also has its own postharvest newsletters, mostly intended for internal communication.

Trade publications. Articles in trade magazines and newspapers are most effective for extending information to industry handlers, who read them regularly. Magazines and bulletins produced by grower-shipper associations include *The Western Grower and Shipper*, published by the Western Growers Association; the monthly *PMA Bulletin*, published by the Produce Marketing Association (PMA); and *Fresh Outlook*, published quarterly by the United Fresh Fruit and Vegetable Association. The weekly newspapers of the fresh produce industry, *The Packer* and *Produce News*, are also included in this group. Articles by extension workers in trade publications extend information to a broad audience that may otherwise be hard to reach. These trade associations and their publications can be very effective media for extending postharvest information. Several postharvest extension specialists have regularly published columns in publications such as these.

Manuals. A practical guide to the use of recommended practices can be developed into a manual for dissemination via sale or distribution as a digital file via the internet. The FAO provides a free manual for "training the trainers" on improving the quality and safety of fresh fruit and vegetables (see fao.org/ag/agn/cdfruits_en/others/docs/manual_completo.pdf). A few of the most popular manuals on postharvest technology can be downloaded free of charge from the UC Davis website, postharvest.ucdavis.edu/libraries/, and a manual on small-scale postharvest handling practices is available in PDF format in twelve languages (postharvest.ucdavis.edu/Library/Postharvest_Center_Publications/). Hundreds of thousands of downloads are made each year.

Visual media. Audiovisual-based training programs and teaching aids are becoming increasingly popular methods of extension as costs of production decline and computer technology makes production and editing feasible on a small scale. UC Agriculture and Natural Resources markets a wide selection of DVDs related to recommended postharvest handling practices. Examples include presentations on postharvest cooling of horticultural crops and postharvest handling of California stone fruit. The UC Postharvest Center and the Postharvest Education Foundation offer free postharvest training videos via YouTube at postharvest.ucdavis.edu/Library/Video_Library/ and via the "PostharvestOrg" channel at youtube.com. The PMA markets several series of videos as training aids for the retail produce industry. Many of these teaching aids can be incorporated into an extension education program. When used as tutorials, audiovisual programs such as PowerPoint presentations, streaming videos, and DVDs can be valuable training tools for short courses and workshops.

Participation in committees and advisory boards

Extension postharvest technology programs can be effectively enhanced through participation in professional society and trade association programs and committees. This means more work for extension personnel, but it can also mean greater program effectiveness. Important professional organizations include the American Society for Horticultural Science and industry trade associations include the United Fresh Fruit and Vegetable Association, PMA, Western Growers Association, International Fresh-Cut Produce Association, Global Cold Chain Alliance, and the Refrigerated Transportation Foundation. Extension personnel should also work closely with committees and advisory boards for fresh-market horticultural crops whenever these have been established by state marketing-order programs. These groups can help extend information more effectively and widely than is possible by working alone.

Developing professional skills

Postharvest extension workers must maintain a high degree of professionalism if the educational programs they plan and implement are to be effective in transferring postharvest technology. It is important to stay current regarding principles and practices of loss assessment and postharvest biology and technology; upgrade skills in extension work, including program planning, extension methods, and evaluation practices; and be highly motivated to identify new problems and assist clientele.

Many tools and skills can assist the extension worker in successfully transferring postharvest technology. In addition to a thorough understanding of the principles and practice of extension methods discussed earlier, the methods of postharvest loss assessment, cost-benefit analysis, and program development are skills of primary importance. These professional activities require direct and regular interaction with various clientele groups. For a new extension worker, finding a mentor among more experienced agents can be a key element in long-term success, since mentors can support, challenge, and provide vision to their protégés (Zimmer and Smith 1992).

Social media

The newest methods being used for professional development are social media or social networking, via websites such as ResearchGate.net and LinkedIn.com, where some of the current open groups include Postharvest Training, Food Value Chains, and the Fresh Produce Industry Discussion Group. While new postharvest professionals may have limited exposure to networks of scientists and to some aspects of practical or field-based postharvest knowledge, and may be unable to participate in conferences or subscribe to many journals due to high costs, most social media websites are free or available at low cost and require only occasional internet access to join in discussions, ask questions of experts, or build a professional network.

Commodity Systems Assessment Methodology (CSAM)

This postharvest loss assessment method sets the stage for productive postharvest extension work by assessing the technical, socioeconomic, cultural, and institutional factors related to handling any given commodity in a specific locale. The end products of CSAM encompass both traditional loss assessment (identifying the causes and sources of losses) and cost-benefit analysis (whether an improved practice will be cost-effective in that particular setting

for that crop) and lead to productive extension program and project development.

The commodity system is made up of twenty-six components that together account for all the steps associated with the production, postharvest handling, and marketing of any given commodity. The method was developed over the course of many years and was tested extensively in the Caribbean before being introduced worldwide to field personnel via an excellent training manual (LaGra 1990). The updated manual includes sample data collection instruments and detailed explanations of each of the components (LaGra et al. 2016). Ideally, teams of people work together while investigating a commodity system—for example, a horticultural production researcher might be teamed up with a marketing specialist and an extension agent. CSAM can help build links between agencies and individuals, close information gaps, and help people solve problems while focusing on usable postharvest technology.

Table 7 shows the list of system components and sample questions for investigating a commodity system. The team begins by considering these questions in relation to any commodity of interest and then adds any other information that is pertinent to the situation. Some of the questions can be answered directly by extension personnel or others who are knowledgeable about the commodity, or information can be found in available literature. Other questions may require the data collection team to observe actual postharvest handling practices and ask questions of those people who harvest, handle, and market the product. Information on the costs and expected benefits of various postharvest technologies can be collected directly or estimated from applied or adaptive research studies.

CSAM helps a postharvest loss assessment team determine the sources of postharvest losses (when and where they occur, and who within the marketing chain is responsible); determine the causes of those losses (what handling or marketing practices are responsible); and estimate the economic value of the losses compared to the costs of current and proposed postharvest practices. Once this kind of information related to training needs has

Table 7. Components of the Commodity Systems Assessment Methodology (CSAM), with sample questions

preproduction		
1.	importance of the crop	What is the relative importance of the crop (number of producers, amount produced, area of production, value)?
2.	government policies	Are there any laws, regulations, incentives, or disincentives related to producing or marketing the crop (e.g., existing price supports or controls, banned pesticides, or residue limits)?
3.	relevant institutions	Are there any organizations involved in projects related to producing or marketing the crop? What are the goals of the projects? How many people are participating?
4.	facilitating services	What services are available to producers and marketers (e.g., credit, inputs, technical advice, subsidies)?
5.	producer/shipper organizations	Are there any producer or marketer organizations involved with the crop? What benefits or services do they provide to participants? At what cost?
6.	environmental conditions	Do local climate, soil, or other factors limit the quality of production? Are the cultivars produced appropriate for the location?
7.	availability of planting materials	Are seeds or planting materials of adequate quality? Can growers obtain adequate supplies when needed?
production		
8.	farmers' general cultural practices	Do any farming practices in use have an effect on produce quality (irrigation, weed control, fertilization practices, field sanitation)?
9.	pests and diseases	Are there any insects, fungi, bacteria, weeds, or other pests present that affect the quality of produce?
10.	preharvest treatments	What kinds of preharvest treatments might affect postharvest quality (use of pesticides, pruning practices, thinning)?
11.	production costs	Estimate the total cost of production (inputs, labor, rent, etc.). What are the costs of any proposed alternative methods?
postharvest		
12.	harvest	When and how is produce harvested? By whom? At what time of day? Why? What sort of containers are used? Is the produce harvested at the proper maturity for the intended market?
13.	grading and inspection	How is produce sorted? By whom? Does value (price) change as quality or size grades change? Do local, regional, or national standards (voluntary or mandatory) exist for inspection? What happens to culled produce?
14.	postharvest treatments	What kinds of postharvest treatments are used? (Describe any curing practices, cleaning, trimming, hot water dips, etc.) Are treatments appropriate for the product?
15.	packaging	How is produce packed for transport and storage? What kinds of packages are used? Are packages appropriate for the product? Can they be reused or recycled?

Continued on page 24

Table 7. Components of the Commodity Systems Assessment Methodology (CSAM), with sample questions, continued

16.	cooling	When and how is produce cooled? To what temperature? Using which method(s)? Are methods appropriate for the product?
17.	storage	Where and for how long is produce stored? In what type of storage facility? Under what conditions (packaging, temperature, relative humidity, physical setting, hygiene, inspections, etc.)?
18.	transport	How and for what distance is produce transported? In what type of vehicle? How many times is produce transported? How is produce loaded and unloaded?
19.	delays/waiting	Are there any delays during handling? How long and under what conditions (temperature, relative humidity, physical setting) does produce wait between steps?
20.	other handling	What other types of handling does the produce undergo? Is there sufficient labor available? Is the labor force well trained for proper handling from harvest through transport? Would alternative handling methods reduce losses? Would these methods require new workers or displace current workers?
21.	agroprocessing	How is produce processed (methods, processing steps) and into what kinds of products? How much value is added? Are sufficient facilities, equipment, fuel, packaging materials, and labor available for processing? Is there consumer demand for processed products?

marketing

22.	market intermediaries	Who are the handlers of the crop between producers and consumers? How long do they have control of produce and how do they handle it? Who is responsible for losses (who suffers financially)? Is produce handled on consignment, marketed via direct sales, or moved through wholesalers?
23.	market information	Do handlers and marketers have access to current prices and volumes in order to plan their marketing strategies? Who does the record keeping? Is information accurate, reliable, timely, and useful to decision makers?
24.	consumer demand	Do consumers have specific preferences for produce sizes, flavors, colors, maturities, quality grades, package types, package sizes, or other characteristics? Are there any signs of unmet demand or oversupply? How do consumers react to the use of postharvest treatments (pesticides, irradiation, coatings, etc.) or certain packaging (plastic, Styrofoam, recyclable)?
25.	exports	Is this commodity produced for export? What are the specific requirements for export (regulations of importing country with respect to grades, packaging, pest control, etc.)?
26.	marketing costs	Estimate the total marketing costs for the crop (inputs and labor for harvest, packaging, grading, transport, storage, processing, etc.). What are the costs for any alternative handling or marketing methods proposed? Do handlers and marketers have access to credit? Are prevailing market interest rates at a level that allows borrowers to repay loans and still make a profit? Is supporting infrastructure adequate (roads, marketing facilities, management skills of staff, communication systems such as telephone, fax, e-mail services)?

been collected, extension educators can target the responsible handlers with appropriate information on cost-effective, improved postharvest technical practices.

In the occasional situation where there is no existing appropriate technical solution for the handling or marketing problem uncovered using CSAM, the problem can be passed on to horticultural researchers in the universities or regional agricultural research centers. The more information provided regarding the commodity system, the better chance the researchers will have to develop solutions that are appropriate to the specific socioeconomic and cultural setting where the postharvest losses occur.

Finally, when institutional or infrastructural issues are determined to be key factors leading to postharvest losses, these can be identified and targeted as advocacy issues. The result of the CSAM process may then be a report making a direct request for policy change or support.

Extension program development

Postharvest extension work requires knowledge and skill in planning, implementing, and evaluating educational programs. Extension workers need to use creativity and initiative in program approaches, teaching techniques, and extension methods. They also must be highly self-motivated and must seek to involve clientele throughout the program. Using informal and formal needs assessments, extension workers must seek out postharvest handling problems and the groups that face those problems. Including clientele in this first stage of program planning is essential, since industry handlers generally will not reveal their problems until extension agents gain their confidence. Encouraging active participation during programs helps people gain maximum value from educational opportunities and increases the likelihood that clientele will provide constructive feedback. For readers interested in more information on this topic, a variety of references related to extension program development are listed at the end of this volume.

Staying current

Extension workers should continue to take courses and short courses in postharvest technology when possible and use sabbatical leaves to advance their capabilities. They must be familiar with modern research techniques, postharvest equipment, and instrumentation. It is expected that each year will bring new ideas and innovations.

In the United States, extension personnel are invited to attend the University of California Postharvest Technology Short Course. In Europe, the Natural Resources Institute regularly offers postharvest training courses. If local extension capabilities in postharvest in-service training are limited, it is possible to work with a variety of development agencies to design and fund training programs. The FAO and the World Bank announce currently available programs on their websites. The U.S. Agency for International Development (USAID) regularly funds postharvest training projects in developing countries through the Farmer-to-Farmer Program and through voluntary agencies such as Agricultural Cooperative Development International/Volunteers in Overseas Cooperative Assistance, in which postharvest extension specialists are invited to participate as speakers and resource persons.

Educational needs assessment

A common complaint raised against extension systems is that programs are related more to the interests of the extension worker than to the needs of the clientele. During the 1980s and 1990s, many evaluations pointed to ineffective extension programs, underutilized postharvest facilities, and disillusioned clientele. In order to avoid this situation, extension workers must regularly use educational or training needs assessment to find out what clients need to learn in order to solve the current postharvest problems they face. Needs assessments may be formal, involving surveys and literature reviews, or they may be informal, based on thorough dialogue with representative clients. Commodity systems assessment is an indirect method for identifying local training needs.

Finding resources

Extension workers will no doubt be required to locate sources of funding for their applied research and extension educational programs. Grant writing skills can be developed by taking courses and seminars on the topic, which are offered at many colleges and universities worldwide. Most libraries have access to the Foundation Directory and the Foundation Grants Index in printed or electronic form. Both of these references contain thousands of entries describing foundations that offer grants to eligible organizations. In addition, the Foundation Yearbook lists over 70,000 foundations that have awarded a cash grant during the most recent fiscal year (foundationcenter.org), and the Catalog of Federal Domestic Assistance lists grants given by U.S. government agencies (cfda.gov/).

It is important to write an effective letter proposal or concept paper when seeking funding for an extension program; much more detailed proposals will be required when applying for most government grants. It can be helpful to ask a trusted colleague to review grant proposals before they are submitted.

Informal teaching and nontechnical writing

Various programs might offer topics specifically aimed, for example, at the problems faced by large-scale growers or shippers of cool-season crops, by small-scale direct marketers of a mixed lot of seasonal fruit and vegetables, or by the fresh-cut industry. Whatever clientele the extension worker intends to serve, the program must be well designed, interestingly presented, and full of useful information. Informal teaching methods and nontechnical writing are tools that can help an extension worker reach a wide range of clientele.

Extension workers who are interested in developing new informal teaching skills or in reading about the successful methods of others can find information in the *Journal of Extension* and the American Society for Horticultural Science's publication *HortTechnology*. Recent articles have dealt with on-farm demonstration, small-group learning activities, task analysis,

and using computers in extension programs. An enormous amount of published and unpublished information on informal teaching methods is available in most educational libraries in the form of microfiche (e.g., ERIC documents).

An entire field is dedicated to agricultural communication, including nontechnical writing for newsletters, extension manuals, fact sheets, and multimedia. Some of the most recent literature on agricultural communication is available from the University of Illinois at Urbana-Champaign and the Illinois Cooperative Extension Service. Back issues and current newsletters on many topics related to extension methods can be accessed free of charge via internet sites such as extension.illinois.edu/global/newsletters. The Department of Human and Community Resource Development at the Ohio State University publishes the quarterly newsletter *Agricultural Communication,* which offers updates on extension methods, writing techniques, and multimedia productions. The editors also review books and short courses dealing with professional skill development.

Marketing extension programs

Even the best postharvest extension programs will be less effective than they could be if people don't know about them. The marketing of extension programs requires planning ahead, developing high-quality promotional materials, and getting the word out via direct mail, posters or brochures, word of mouth, or the internet. The four P's of marketing (product, place, promotion, and price) are also important for marketing postharvest extension programs (table 8). As with quality program planning, where the needs of the clientele determine the objectives of the extension effort, the more an extension worker knows about the target audience, the more likely it is that the program will be marketed to those who will benefit most from participation.

Program evaluation is the last subject included in this topic of postharvest extension and capacity building. Extension workers must be willing to do the work associated with evaluating their ongoing projects and

Table 8. The marketing mix

product	What will the postharvest extension program involve? What new information is being offered to clientele?
place	How and where will the program be delivered to growers, shippers, and/or marketers?
promotion	How will the planned target audience be informed about the program and encouraged to participate?
price	What costs or fees are part of the extension program?

Source: Adapted from Brown 1984.

extension programs in order to determine what is working well and what needs to be changed.

Evaluation requires gathering information during implementation to determine whether the program has met its objectives. Bennett's model (1979) describes seven levels of evidence related to a program's theory of action, which can help an extension worker focus the evaluation on the key aspects of the program: inputs, activities, participation, reactions, knowledge/skill changes, practice/behavioral changes, and end results. A theory of action attempts to understand causal linkages between the program inputs and activities and the program's intended outcomes. It is important to focus attention and resources during evaluation on higher-level outcomes, such as actual or intended practice changes, rather than counting the number of participants and assuming they learned and used the postharvest technologies offered during the program.

In summary, many extension methods have been used successfully to assist clientele to identify key problems and assess potential solutions for their technical and economic feasibility. The goal of extension is to make the link between those who are developing new postharvest technologies and those who are interested in adopting them. Postharvest extension specialists and extension workers worldwide have the opportunity to identify and transfer appropriate postharvest technologies that will assist their clientele to reduce produce losses, better maintain quality and value, and, most important, increase profits during handling and marketing.

Case study: Postharvest extension in California

The UC Postharvest Center, formerly known as the Postharvest Outreach Program and then the Postharvest Technology Research and Information Center, is a research and extension center where UC Cooperative Extension specialists work collaboratively with industry partners to perform research studies and provide educational programs in the interdisciplinary field of postharvest technology of horticultural crops. Members of the program are based at the campuses of UC Davis and UC Riverside and at the Kearney Agricultural Center in Parlier. The specialists represent several academic departments, including plant sciences, food sciences, agricultural economics, and biological and agricultural engineering. A wide range of laboratories and controlled-temperature chambers are available for applied postharvest research and demonstration activities. The Center cooperates closely with a variety of organizations representing the horticultural industry and seeks assistance, as required, from emeritus extension specialists and private California-based postharvest consultants.

Purpose of the program

The UC Postharvest Center's program focuses on particular problems the industry faces in handling fruit, vegetables, and ornamental crops. California growers supply nearly half of the fresh fruit and vegetables consumed in the United States. They also supply millions of cut flowers and other ornamentals that contribute to quality of life. The program's goals are to improve the quality and value of horticultural crops available to the consumer, reduce postharvest losses, and improve marketing efficiency.

Program development and management

Through the postharvest research and extension leadership efforts of the extension specialists, the program strives to provide relevant information for all California growers, shippers, marketers, carriers, distributors, retailers, processors, and consumers of horticultural crops. Members meet regularly to discuss research results and program evaluations and to plan a variety of research and educational activities in response to industry needs. These activities include

- problem-solving research related to maintaining the quality and safety of horticultural produce between harvest and consumption

- short courses, workshops, and seminars on topics related to postharvest technology

- publications such as textbooks, manuals, produce fact sheets, and newsletters

- audiovisual materials such as slide sets, PowerPoint presentations, and videos

- industry meetings where program members present research results and postharvest information

- consultations by telephone and in person with individual handlers

Several publications and videos have been translated into Spanish and other languages. Outreach activities include a website, postharvest.ucdavis.edu, containing a calendar of events, information on the program, and links to a wide selection of related websites. Occasionally, faculty members are invited to plan and implement short courses and workshops in foreign countries.

Resources

The availability and sources of funding for the program differ widely from year to year. At any given time, funding for research and educational programs may come from the following:

- competitive grants from state and federal government agencies, industry sources, university funds, or private foundations

- participant fees from short courses, seminars, and workshops

- sales of postharvest manuals, educational materials, and back issues of newsletters

- grants from UC administration funding sources for support of UC Research and Information Centers

- support for specific projects from organizations such as USAID, the World Bank, and the FAO

- interest income from the UC Postharvest Program Endowment fund, established to support teaching, research, and extension activities related to postharvest biology and technology of fruit, vegetables, and ornamental crops

Successes

The collaborative efforts of the UC Postharvest Center have resulted in the development of several postharvest handling practices that are now used extensively throughout the horticultural industry to reduce losses and maintain produce quality. Tight-fill packing was introduced in the 1960s, and a forced-air cooling system (originally designed by Rene Guillou in the 1950s) was adapted for use with cut flowers during the 1970s and is now the industry standard for many highly perishable crops. The elucidation of the ethylene biosynthesis pathway during the 1980s led to the development of methods for manipulating the ripening of fruit and to the increasing use of commercial ripening facilities. When the horticulture industry in California moved into the fresh-cut produce sector, the Center added applied research and outreach activities aimed at maintaining quality and fresh-cut product safety. The leaders of the Center have been involved in the management of the USAID Horticulture Innovation Lab, which funds a series of horticultural development projects in Africa, South Asia, Southeast Asia, and Latin America.

The educational activities of the UC Postharvest Center reach hundreds of people annually in short courses and workshops, and thousands of texts and manuals have been distributed. Each year several new titles are added to the collection of postharvest extension publications, and older publications, slide sets, and videos are regularly updated. The Center website has been recently upgraded and receives over a million page views each year.

Industry partners as well as the entire horticultural community benefit from their collaboration with the Center. Their mutual goals of reducing losses, ensuring safety and freshness of horticultural products, and improving marketing efficiency are critical to maintaining California's reputation for produce quality.

Constraints

When considering the UC Postharvest Center as a model for postharvest extension efforts, it is useful to note just a few weaknesses in the program. The relative autonomy of the faculty postharvest specialists and their diverse locations around the state of California sometimes make it difficult to accomplish planned tasks. In the absence of a strong leader, many projects would tend to stagnate. Much of the success of the Center therefore is due to an individual leader's dedication. He or she must spend much time and energy motivating members and overseeing projects to ensure their timely completion.

The lack of reliable funding is another weakness. The program director and other members require strong fundraising skills in order to complete their applied research, outreach, and service objectives. Planning long-term activities is especially difficult due to the variation of funding sources and amounts that may be available each year.

References

Adhikarya, R. 1994. Strategic extension campaign: A participatory-oriented method of agricultural extension (a case study of UN FAO's experiences). Rome: United Nations Food and Agriculture Organization.

Andrew, C., and P. Hildebrand. 1993. Planning and conducting applied agricultural research. Boulder, CO: Westview Press.

Barao, S. 1992. Behavioral aspects of technology adoption: The role of on-farm demonstration. Journal of Extension 30 (summer):13–15.

Bennett, C. 1979. Analyzing impacts of extension programs. Washington, D.C.: USDA Extension Service.

Blackburn, D., ed. 1994. Extension handbook: Processes and practices. Toronto: Thompson Educational Publishing.

Breslin, P. 1996. Costa Rican farmers find their mini-niche. Grassroots Development 20(2):27–33.

Cantwell, M. 2007. Postharvest handling of horticultural products: Keeping principles in perspective. Small Farm News 1.

Cernea, M. 1991a. Putting people first: Sociological variables in rural development. 2nd ed. New York: Oxford University Press.

———. 1991b. Using knowledge from social science in development projects. Washington, D.C.: World Bank Discussion Papers no. 114.

Chambers, R. 1991. Shortcut and participatory methods for gaining social information for projects. In M. Cernea, ed., Putting people first: Sociological variables in rural development. 2nd ed. New York: Oxford University Press. 515–537.

Chambers, R., A. Pacey, and L. Thrupp. 1989. Farmer first: Farmer innovation and agricultural research. London: Intermediate Technology Publications.

Chinsman, B., and Y. S. Fiagan. 1987. Postharvest technologies of root and tuber crops in Africa. In E. Terry, et al., eds., Tropical root crops: Root crops and the African food crisis. Proceedings of the 3rd Triennial Symposium of the International Society for Tropical Root Crops, Africa Branch. Owerri, Nigeria: ISTRC.

Compton, J. 1997. Problems, technologies and participation. International Agricultural Development 17(4):21.

Ewell, P. 1990. Links between on-farm research and extension in nine countries. In D. Kaimowitz, ed., Making the link: Agricultural research and technology transfer in developing countries. Boulder: Westview Press. 151–196.

FAO (United Nations Food and Agriculture Organization). 1985. Prevention of post-harvest food losses: A training manual. Rome: FAO Training Series 10.

———. 1989a. Horticultural marketing: A resource and training manual for extension officers. Rome: FAO Agricultural Services Bulletin 76.

———. 1989b. Prevention of food losses: Fruit, vegetable and root crops. A training manual. Rome: FAO.

Gustavsson, J., C. Cederberg, U. Sonesson, R. van Otterdijk, and A. Meybeck. 2011. Global food losses and food waste: Extent, causes and prevention. Rome: FAO.

IFPRI (International Food Policy Research Institute). 1995. Report highlights barriers to women. International Agricultural Development 15(5):4.

Kader, A. 2005. Increasing food availability by reducing postharvest losses of fresh produce. Acta Horticulturae 682:2169–2175. https://doi.org/10.17660/ActaHortic.2005.682.296

———. 2006. The return on investment in postharvest technology for assuring quality and safety of horticultural crops. Journal of Agricultural Investment 4:45–52.

———. 2010. Handling of horticultural perishables in developing vs. developed countries. Acta Horticulturae 877:121–126. https://doi.org/10.17660/ActaHortic.2010.877.8

Kader, A., and R. Rolle. 2004. The role of post-harvest management in assuring the quality and safety of horticultural produce. Rome: FAO Agricultural Services Bulletin 152.

Kader, A., L. Kitinoja, A. Hussein, O. Abdin, A. Jabarin, and O. Sidahmed. 2012. Role of agro-industry in reducing food losses in the Middle East and North Africa Region. Rome: FAO.

Kitinoja, L. 1999. Costs and benefits of fresh handling practices. In L. Kitinoja, ed., Perishables Handling Quarterly special issue: Costs and benefits of postharvest technologies. No. 97:7–13.

———. 2010. Identification of appropriate postharvest technologies for improving market access and incomes for small horticultural farmers in Sub-Saharan Africa and South Asia. WFLO Grant Final Report to the Bill & Melinda Gates Foundation. http://ucce.ucdavis.edu/files/datastore/234-1847.pdf.

———. 2013. Innovative small-scale postharvest technologies for reducing losses in horticultural crops. Ethiopian Journal of Applied Science and Technology special issue 1:9–15.

Kitinoja, L., and D. Barrett. 2015. Extension of small-scale postharvest horticulture technologies—A model training and services center. Agriculture 5 (2015):441–455. https://doi.org/10.3390/agriculture5030441

Kitinoja, L., and J. Gorny. 1999. Small-scale postharvest technology: Economic opportunities, quality and food safety. Davis: University of California Postharvest Horticulture Series 21.

Kitinoja, L., and A. Kader. 2003. Small-scale postharvest practices: A manual for horticultural crops. 4th ed. Davis: University of California, Davis. http://ucce.ucdavis.edu/files/datastore/234-1450.pdf.

Kitinoja, L., and J. Thompson. 2010. Pre-cooling systems for small-scale producers. Stewart Postharvest Review 6(2):1–14. https://doi.org/10.2212/spr.2010.2.2

Kitinoja, L., S. Saran, S. Roy, and A. Kader. 2011. Postharvest technology for developing countries: Challenges and opportunities in research, outreach and advocacy. Journal of the Science of Food and Agriculture 91(2011):597–603. https://doi.org/10.1002/jsfa.4295

Krimsky, S., and A. Plough. 1988. Environmental hazards: Communicating risks as a social process. Dover, MA: Auburn House.

LaGra J. 1990. A commodity system assessment methodology for problem and project identification. Moscow, ID: Postharvest Institute for Perishables. http://www.fao.org/wairdocs/x5405e/x5405e00.htm.

LaGra, J., L. Kitinoja, and K. Alpizar. 2016. Commodity systems assessment methodology for value chain problem and project identification: A first step in food loss reduction. San Jose, Costa Rica: IICA.

Lipinski, B., C. Hanson, J. Lomax, L. Kitinoja, R. Waite, and T. Searchinger. 2013. WRI working paper: Creating a sustainable food future reducing food loss and waste. World Resources Institute. http://pdf.wri.org/reducing_food_loss_and_waste.pdf.

Madeley, J. 1987. The African farmer ... and her husband. International Agriculture and Development 7(2):1.

Manalili, N., M. Dorado, and R. Van Otterdijk. 2011. Appropriate food packaging solutions for developing countries. Rome: FAO.

Narayanan, A. 1991. Enhancing farmers' income through extension services for agricultural marketing. In W. Rivera and D. Gustafson, eds., Agricultural extension: Worldwide institutional evolution and forces for change. Amsterdam: Elsevier. 151–162.

Patil, R. 2010. Appropriate engineering and technology interventions in horticulture for enhanced profitability and reduction in postharvest losses. Acta Horticulturae 877:1363–1369. https://doi.org/10.17660/ActaHortic.2010.877.187

Rhoades, R. 1984. Breaking new ground: Agricultural anthropology. Lima, Peru: International Potato Center.

Rogers, E. 1995. Diffusion of innovations. 4th ed. New York: Free Press.

Saran, S., S. Roy, and L. Kitinoja. 2012. Appropriate postharvest technologies for improving market access and incomes for small horticultural farmers in subsaharan Africa and South Asia. Part 2: Field trial results and identification of research needs for selected crops. Acta Horticulturae 934:41–52.

Schurr, C. 1988. Packaging for fruits, vegetables and root crops. Bridgetown, Barbados: FAO.

Shewfelt, R., and S. Prussia, eds. 1993. Postharvest handling: A systems approach. San Diego: Academic Press.

Stuart, T. 2009. Waste: Uncovering the global food scandal. New York: Norton.

Van der Ban, A., and H. Hawkins. 1996. Agricultural extension. 2nd ed. Oxford: Blackwell Science.

Vermeulen, S., J. Woodhill, F. Proctor, and R. Delnoye. 2008. Chain-wide learning for inclusive agrifood market development: A guide to multi-stakeholder processes for linking small-scale producers with modern markets. London UK: IIED; Wageningen, the Netherlands: Wageningen University Research Center.

Winrock International. 2009. Empowering agriculture: Energy options for horticulture. US Agency for International Development. http://ucce.ucdavis.edu/files/datastore/234-1386.pdf.

Zimmer, B., and K. Smith. 1992. Successful mentoring for new agents: Dedicated mentors make the difference. Journal of Extension 30 (spring): 25–27.

Index

W

Y

Z

CPSIA information can be obtained
at www.ICGtesting.com
Printed in the USA
JSHW010051111122
32836JS00005B/5